craftsmen of necessity

CHRISTOPHER WILLIAMS
photographs by Charlotte Williams

Random House New York

Craftsmen OF Necessity

Copyright © 1974 by Christopher Williams
All rights reserved under International and Pan-
American Copyright Conventions.
Published in the United States by Random House,
Inc., New York, and simultaneously in Canada by
Random House of Canada Limited, Toronto.
Portions of this book have appeared in *Shelter* and
Industrial Design Magazine. An article entitled "Crafts-
men of Necessity" by Christopher Williams which con-
tains some of the material in this book appeared in
the November, 1972 issue of *Natural History* Maga-
zine. Copyright © 1972 by The American Museum of
Natural History.

Library of Congress Cataloging in Publication Data
Williams, Christopher G
  Craftsmen of necessity.
  1. Technology. 2. Human ecology. 3. Dwellings.
I. Williams, Charlotte E., illus. II. Title.
T49.5.W53 1974 600 73-19692
ISBN 0-394-48983-7

Manufactured in the United States of America
98765432
First Edition
Designed by J. K. Lambert

When man first began to build, he watched nature's performance and followed it. His ways rested easily in the environment because the environment was his control and reference; he lacked the power to do otherwise. From the surroundings came his materials: the plants and animals, the rock and dirt. He converted the materials into his tools. He learned the vernacular of his materials, the strength of wood, the shapes of clay, the cleavage of stone. He learned the pressures and demands of the biosphere, and he bent to them.

The builders who worked wood were careful to allow for the intricacies of the tree's growth, to take advantage of grain and limb, and the tree also shaped the design. The farmers' plows were made of wood and leather and were designed to fit the nature of each particular type of soil lest they break or the limited animal power fail. Environment shaped men's dwellings and their lives. Over the countryside people assembled their own order, apart from nature but closely associated to its temperament and mood.

With their simple technology this association remained harmonious, but as machine technology gathered its strength, man followed it. The designing and building of man's personal environment were taken from individual hands, to be placed under the remote decisions of technology. Machine technology eliminated the personal designers, and building left the home to go to the specialists and the factory. Back came the goods of general production, glistening machines of precision the human hand could never duplicate.

Machine technology drove a steel wedge between man and his home. The farmer who was on intimate terms with his land grew to know it more remotely, for a steel plow need not take account of the peculiarities of regional dirt as it slices through the sod. Mechanical woodcutting is not concerned with deviations in a tree's growth; it tries to reduce everything to its simplest form. Technology has given builders materials and methods that have little to do with climate or environment. Man has left nature's domain for his own. The land has

3

become only a platform to hold his houses and cities and to be manipulated into growing his produce.

Man has come full circle. He has risen from a unit of the biosphere to become an ecological force himself. But the technology that has elevated him is a crude and artless power. It manipulates with overforce, it is graceless and inefficient, complicated in its manifestations but simplistic in its concepts, subject to failure and collapse. Its energies are hostile, vulgar and overextended; it reproduces itself with tiresome repetition and inspires a dull, unimaginative human life.

Today machine society is still not universal. Another half of the world lives without modern technology. Its inhabitants have concepts of form and function that have been developed out of the biology of their land. They live and work within the restriction of their country and climate.

Western society chose to embrace the machine. This choice is still not open to most of the indigenous builders of the world. But as economic situations change, modern technology becomes available to them, too, and like the West, they close their eyes to the disciplines of the living world and heedlessly leap into the mechanical technology of the twenty-first century. The pressures of their environment lift, and they become perpetrators of environmental crimes far worse than our own. But is it possible to live with the environment and alter it, as man must do, and still not violate it?

Most of the indigenous people of the world still practice organic technology. This is the opposite of machine thinking. It is a way, not a device, a philosophy to govern the methods of selecting action. It does not plan for an event but lets the event develop on its own and then coordinates its effort with the development. It is perceptive of environment both human and natural, for with organic technology the two are not at odds. It is unlimited in its growth, and as it grows it slips from one system to the next rather than overextending and complicating limited systems. It creates natural cycles that balance themselves to reuse efforts, energies and materials; waste is absorbed and recirculated. Organic technology never has the stamp of duplication, every product is individual. Materials are used with the method and shape best suited to the particular structure of each, for organic technology is above all honest—to itself and to those who work under it.

The following pages examine people, towns, mud, forest and stone and attempt to give substance to these perilously broad definitions.

Those who live with nature are forced to ride the currents of ecological fortune. For this reason their ways are important to us who manipulate. This book should not be interpreted as a nostalgic backward glance. Its intent is to glean the essence of a particular quality that they have and we do not, that we have forgotten to take with us.

A family feels its smallness when it is in the center of a world that ripples over eroded desert mountains to the horizon on four sides without a break. The indigenous builder's work must stand alone as a buffer between the people and the environment. With pride and confidence he manipulates the dusty soil under his feet to build a home that contains the warmth of his family and moderates the heat of the sun. The people who live in the least giving and most demanding environment must reply exactly to its requirements. The indigenous builder's ability flowers in the hostile climates of the world.

~

In southern Tunisia, at the edge of the Sahara Desert, a small group of people have sustained a stable life inside the ground. In elaborate hollows, wells and tunnels they have maintained their society. The village of Tijia in the mountains southwest of Gabes contains two hundred people beneath the ground.

Where the temperature is an inferno at noon and drops precipitately at night, where there are no trees for building, no reeds and rocks, no mud or animal skins, no alternative, the people dig into the earth to live; they practice the art of subtractive building.

Just beyond the palm trees on

the far slope are four deep shafts in the hillside. They form the central courtyard of four homes. The village lies tucked into these low hills so completely that the unknowing traveler can pass within a few hundred feet of the town center and not detect it. The people and the town they have built are absorbed visually into the desert, because the desert has offered them little to use and nothing to squander. Nothing rises above the desert's horizon, and only a few small worn paths connect the unobtrusive burrows. There is no debris from human activities, no garbage, no scatter, no waste, for there is nothing to waste, there is nothing left over.

A man from this country intending to build a home for himself and family selects a soft clay hillock, perhaps fifty feet from base to top. Before he begins he knows that there, in the dirt of the hillside, is the complex of rooms and halls all put together. The doorways are framed, the rooms have ceilings and walls, closets and crannies. It is up to his imagination to find where they are; up to his hands to remove the material

7

that surrounds them. In the architecture of subtraction, the shape things take is not controlled by arbitrary configurations; size and shape are regulated by the composition of the material and the weight from above. With shovel and mattock the builder digs a large shaft into the top of the hill,

thirty feet in diameter and forty feet deep. The base of this shaft forms a courtyard. It provides light and air and a connection between the rooms. The hill's interior is regarded as three-dimensional space. The house can expand in any direction, cistern and storage wells can go down, and there is the height of the hill for upward expansion. Often the apartments are stacked two or three stories up the central shaft.

The entrance to the house is at ground level, outside the hill at its base, and leads to the courtyard floor, which is the same level. The rooms—for sleeping, cooking, storage and livestock—radiate off this courtyard. The doorways are carved to accommodate the amount of traffic and the human form. The larger ones must be vaulted to hold the weight from above. Children's quarters are often higher on the shaft wall and smaller; the livestock stalls are large and round. The inhabitants never regard the walls of their home as fixed, but rather as a plastic median. New openings and passageways are cut as the occasion arises.

The motions of the builder's hand are written on the walls of his home in little arched cuts. His efforts and personality are built into the house. With the years, the house will mold to fit the family. Rooms spontaneously sprout off others as children are added, and are filled in when they leave. Doorways and windows are made by the hand of a child: nooks and crannies of various shapes blossom to hold candles and small precious things. Through time the house undulates with the movement and size of the family, and always the builder lives with and can contemplate his work.

To the right is the kitchen, consisting of several rounded hollows. The smaller act as ovens, the larger are for storage and food preparation. Cooking is managed with a very small fire, just enough for the job required, for fuel also is scarce; most of it comes from dried animal dung and occasional charcoal. The few pieces of wood used in building are from the treasured palm trees. These are seldom cut while alive, so the woodworker must wait until one

dies. Pots, pans and utensils come from the distant markets and are purchased only one in a lifetime. High on the wall of the shaft, pigeons nest in the niches cut for them. With their yield of meat, eggs, feathers, bones and fertilizer, the pigeons are an integral part of the economy.

These people eat what they can. They shelter goats in the courtyards of the central well and occasionally a sheep. For a month during early spring clumps of spiny grass grow in the low places; these they harvest for the sheep, the goats must forage for themselves. The goats provide milk, meat, leather and bone; the sheep, clothes and blankets; the earth, dyes; the few palm trees, fronds for reedwork and dates. There is no cushion between the supply available and the minimum to be consumed, a closed and very tight system.

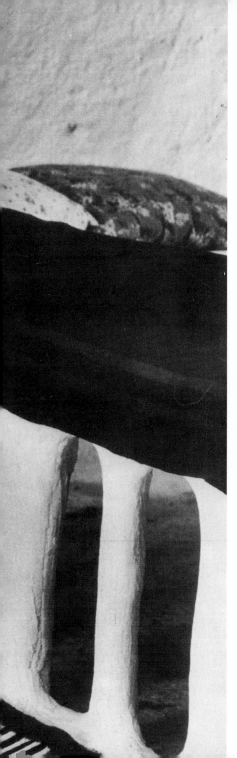

Conventional architects add boards and beams, bricks, cement, as an accountant would add a column of figures to arrive at a sum. The subtractive architect cuts and chisels into the existing form, looking for the negative spaces in which he will live. In some rooms that he knows will be used for sleeping, eating or storage he leaves blocks and lumps projecting into the interior space. When the apartment is finished, the builder comes back to these projections and carves them into beds, bins, shelves and cupboards, which flower from the walls and floor.

The blankets and rugs in photograph 6 were woven in this room, of wool carded and spun in the courtyard.

~

A few miles deeper into the range of the Matmatas, the climate is the same, but the environment has allowed the inhabitants the luxury of another material: stone.

A village of Sahara Berbers has established a permanent home on a mountain of rubble stone. Stone houses them, holds their water,

food, stairways and latrines, grinds their meal and cooks their bread. The village lies within the upper section of the peak in the background. Aside from the traditional Moslem church, all the houses are backed into the mountain. Some are caves cut into the mountainside, others project up and out. The mountain is encircled with level upon level of stone paths, each supplying a row of dwellings. Some of these are permanent, others have been rebuilt hundreds of times over centuries. The town is one of the oldest continuously inhabited in the Matmatas.

In the foreground is the retaining wall of a damp soil reservoir. There are several of these in ravines and lower valleys of the mountain. The spring rains wash the few particles of soil off the mountainside; the retaining walls stop the soil and let the excess water pass. Over the years enough dirt is accumulated to sustain growth. Grain is grown in the smaller reservoirs higher on the mountain; the larger ones at the bottom hold enough deep moisture for palm and olive trees to grow.

Most of the abundant stones can be handled by one man. They are angular and uniform in size and shape and considered the common property of the inhabitants of the village. The villagers regard the stones as a reusable element. As a result, the town is always changing its form, although the population is stable. An unused building or a pile of loose stones is readily available material for a family wishing to build. Houses appear without plan; the ledges, slopes, paths and adjacent buildings are their guide. There is never any waste because the material is considered a permanent element; the form it assumes is used, then discarded.

Four thousand miles to the north of Tunisia, above the Arctic Circle in the high tundra of Lapland, circumstances have forced the Lapps to become transients, moving back and forth over the bleak land with the reindeer on which their livelihood depends. With them they carry sapling poles cut from groves of stunted birch, the only tree strong enough to survive this far north. They bend the poles into bows, then cross several in their centers and tie them. These form the skeletal structures of their homes. Over them the Lapps throw reindeer skins and sod to form small domed mounds, thick-walled against the cold. When they move to a new location, the poles and skins are tied together; the sod is replaced.

Both peoples, the Lapps and the Berbers, live in barren environments. Both must cope with difficult temperatures and limited materials. Both are products of environmental pressures.

egypt's arable land is a thin green sliver of vegetation around the Nile River, pointing north from equatorial Africa through the white desert to the Mediterranean Sea. On this narrow vein of greenery four hundred miles long the population must live and work. The brown and green fields extend up a hill on each side of the Nile Valley, then end as abruptly as a shoreline at the edge of the dry white sand. Though the land is fertile, the desert is always visible, beyond the edge of the field. The Egyptians are never allowed to forget that they are surrounded by a sea of dryness.

The Egyptians' life is not balanced on the sharp edge of supply and demand as it is in the desert, but environmental pressure is still great enough to force them to use everything available; there is no waste.

Their environment is composed of four primary elements: the sand, the river, the waterborne mud carried by the ebb and flow of the Nile, the palm tree that is the most important thing that grows.

The Nile's long banks have been tended and worked more intensely for more years than any other possession of mankind. The river is a continuum from the past in culture and work, and it exists as completely for the Egyptians today as it did 3,000 years ago.

The Nile Delta may be among the few places in the world where the natural environment has profited from man's actions. Although the land is flat, the river passes a few feet below its surface and much of it would remain dry if unbroken generations of Delta people had not cut and maintained ditches and troughs. Through heavy wooden-geared animal-powered water wheels and endless chains powered by peasants' feet, they raised it to the surface from wells.

~

If the fields to be cultivated lie close to the open water, the ancient shaduf is used to raise the water to field level. The counterpoise lifts the filled bucket the necessary· distance to save the man's back. The man in the following photograph can move about 800 gallons a day. When the dis-

tance is only two or three vertical feet from one irrigation ditch to another, the so-called Archimedes screw is used (see photograph at right). This spirals the water up the helical inside of the drum to spill out on the other side. History ascribes this invention to Archimedes, although in fact the Egyptians were using it long before he was born.

The recently completed High Dam of Aswan, constructed to produce electricity for Egypt's industrial needs and to create a lake in an attempt to establish fertile conditions in the arid land of Upper Egypt, has had severe, adverse effects on the rural people of the lower Nile Valley. Before the dam, there was an annual inundation of the delta fields, which came at a time when there were no crops down and so did little damage and kept the fields supplied with fertile soil. Today the soil is less able to produce because the fields no longer have this annual influx of mud. The fish at the delta mouth have become scarce and in some places have disappeared, because the river no longer maintains food supply. And an infection carried by snails, which attacks the human body, has reached epidemic proportions as the snails, which prefer still water, have proliferated.

Egypt's indigenous industry rests on two elements in the Nile environment: mud and the palm tree. Mud from the river has become both farmland soil and building material for houses, roadways, ovens, cooking utensils, reservoirs, piping, roofing, toys, even furniture. Scooped from the riverbank and mixed with chopped straw, the mud is cast in bricks or tramped and compressed into wooden forms that slide up as walls to form a thickly sheathed town. The house in the background in the photograph on the right is built of precast mud brick covered with a coating of mud. Fresh bricks are stacked in the dooryard.

~

Building materials are always readily available. The bread oven above was finished just a few minutes before the picture was taken. It was built by the woman who intends to use it when it is ready for firing. The design is her own, and to complete it she hastily added the finger decoration. The Delta people look to themselves with complete confidence to solve their own problems.

Kitchen crockery and small items are made from refined mud. Through a series of canals and ditches the river water is channeled to large evaporating pools. From one to the next the mud slowly thickens as the sun leaches out the water. It goes into the final pool as heavy brown syrup and there changes from liquid to solid. The soft mud is cut like brownies in a pan, then removed in large squares. Clean and refined, it is now ready for the potter's wheel.

The local potter supplies all the wares the townspeople are unable to make themselves. His production runs from delicate jewelry boxes to sewer pipe.

~

One of the important products of the Egyptian potters' industry is the water jug. The hollow beehive-shaped bottoms are made first, the tops are added after the clay of the bottoms has become firm, and the completed jug is left to dry in the shade before it is fired. The firing is done in a large clay furnace, which is rebuilt each time. When enough pieces have accumulated, they are piled together with twigs, branches and animal dung to a height of twenty feet, then covered with a cone-shaped wall of clay. A few air holes are made at the bottom and top. The furnace is lit and allowed to burn slowly for several days; it is then torn apart to remove the bisque pottery from inside.

Water jugs are left unglazed. The natural porosity of the clay allows the water to soak through the vessel and dampen the exterior. The slow evaporation from the outside of the jug cools the water inside.

Towns along the Nile are built of either mud and straw bricks, or of mud spread over a palm-frond thatch, like the wall behind the three boys in the bottom photograph at the far left. Space is valuable, so most houses are compressed into villages, which are composed of vertical and horizontal structures carefully fitted together.

Such villages grow spontaneously; an outside wall suddenly becomes the inside wall of a new apartment; shared walls support individual and mutual roofs, roofs become footing for other apartments and storage for straw. The town is on a series of levels, so privacy is maintained. Trees are left standing for shade; the buildings mold around them. All the work is done by the inhabitants. The construction is simple, the material available, and when there is a change to be made, the inhabitants are free to act on it immediately without depending upon outside help. They are in total control of their personal environment. Consequently, the work is accomplished quickly and easily. The buildings are in a continual process of evolution.

The long serpentine trunks of the palm tree rise above every village and about every field. The fibrous palm has entered almost every facet of the people's lives. It is their first line of defense against the sun in the open fields, and in its shade grows the olive tree. Under the olive, the fig grows, and under the fig, the pomegranate and vine, then the grain and vegetables. The palm tree's second contribution is dates, which in turn produce a luxuriance of foods; its third is its long flexible fronds. These are cut, punched, split, chopped and shredded. Their heavy bases are used as laths for buildings, their centers for the crates in which most of Egypt's merchandise and small livestock move.

The ends of the fronds are woven into mats or split into smaller sections for baskets, rugs and sandals. Splitting them still fur-

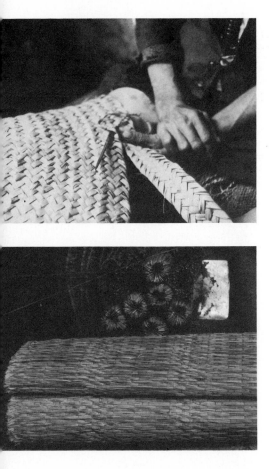

ther produces a kind of straw that is used for brushes and brooms. Or they may be shredded into a fine material that can be braided into twine and then twisted into rope. The very fine tips are generally used for hats.

The date palm must be fertilized to produce fruit, which has given rise to the occupation of palm seeder. In his shop the seeder has a selection of male seeds with selected pods (see photograph at right). Standing on a ladder, he spreads these on the branches of the female tree for the next season's dates. When the palm grows old, its trunk is used for lumber. It is poor construction material but adequate in large sections. It is used mainly for roof beams that cross horizontally from one wall to another. The palm's final yield is the palm liquor taken from the very heart of the tree when it is first cut.

Each part of the tree is used; not a fiber is wasted. The root ball is burned for cooking, and from its ashes soap is made.

The materials of the earth are finite and so must be constantly broken down for reuse in other forms. The earth is a vast closed system, with only the introduction of sunlight to catalyze organic action. The biosphere produces individuals in vast quantities in order that a few may be certain to reach maturity and produce again.

At first view the biosphere appears to be a great green organism of production and waste, giving birth to vegetables and animals that are consumed by disease and decay, fail to reach fertile ground or are pressed out by competition, but because it is a closed system, there is no waste, for with a finite operation, there can be no linear consumption. The multitude of delicately produced seeds or eggs that are never given the chance to multiply fall back into the organic debris or are taken in as high-energy food by a more successful individual. The only input that is totally consumed is sunlight. All else is broken down for reuse.

The system is new in galactic terms. There are indications that earth materials have not yet found an equilibrium. Universal and earth systems recirculate rather than waste because of the pressure to economize on matter and energy. Indigenous builders, too, are forced by the need to economize on means and material to use all that is available to them. Only this scarcity keeps them within the system.

When a human society has been living in the same place for a long period of time and operating in the same manner without the need to change because of diminishing supplies, we can assume that that society is in balance with its environment. It consumes, but not enough to deplete; its wastes and excesses are somehow utilized.

The few human societies that are apparently tolerated by their environment tread lightly, not because they want to, but because they must. As environment controls the population and activities of the other forms of life, it also controls the indigenous societies of man. They seek and find methods of working that do not violate their surroundings because they lack the power to survive if they violate. They use self-regulating, recirculating systems because they do not have either the power or the material to squander. They compromise because they must.

In their building, indigenous societies often use materials and forms the way they have been created by the environment as a result of the forces of wind, water or earth movement. They are forced to do this by necessity, but the result is more than the product of need.

~

The indigenous builder responds to his environment in all its aspects. The climate determines the extent to which he has to shelter his family; the terrain suggests the forms his building should take; the stones, dirt and vegetation provide his materials.

On the northwestern plains of Syria the rainfall is adequate for grazing but sustains no trees or bushes, and there is little rock. The descendants of Bedouins who live here build mud houses that are smooth, rounded mounds called "beehives" by local people. Their shapes are influenced mostly by the consistency of the mud.

When a soft yielding material like mud or clay is used with another, harder, stronger material like wood, the rigid material dominates the shape. The Egyptians' village houses are built with clay walls, but wood beams cross the ceiling, and the house takes the geometric form of a post-and-lintel building. The walls rise straight and the roofs are flat; the doors and windows are square because that is the best way to join two flat straight members at a corner. Syrian villagers have only clay available and must make do with it. They tamp and compress it into a vertical wall, but without a strengthening member the only way for the mud to span the opening between the walls is with a dome. In these villages the mud has been left to find its own form and has molded the inhabitants' lives into a small world of curved plumpness.

In some villages one can see the encroachment of new industrialized housing, standing out square and rigid against the rounded forms. Materials for the flat-roofed buildings were of course imported. The domed housing is modular, since the maximum size of the dome is limited. A single-family house may consist of several domes; some will be sleeping quarters, others kitchen, storage, livestock and working space. The small rounded tip at the peak of the dome is decoration and also acts as a reinforcement to absorb the erosion of the weather.

The photograph on the left shows the severe contrast between the dictates of construction with stone and with soft forms of mud. A stairway becomes a smooth, even random grouping of levels when made in the soft material. Mud squares up only when it has to meet a rigid material, like the wooden-framed doorways. The dome in the center is a main sitting room; to the right is a storage hut.

When windows and doors are left unframed, they assume their own soft, round openings. The town is a delight to be in; it it is a labyrinth of winding, intricate many-leveled walks and alleys where walls slope in and out, doorways are rounded openings and the scale constantly changes. In courtyards like the one in the photograph above, the scale is very small—a large man would have difficulty passing through the doorway. Then, as the passage opens into a broad street, the scale also shoots up.

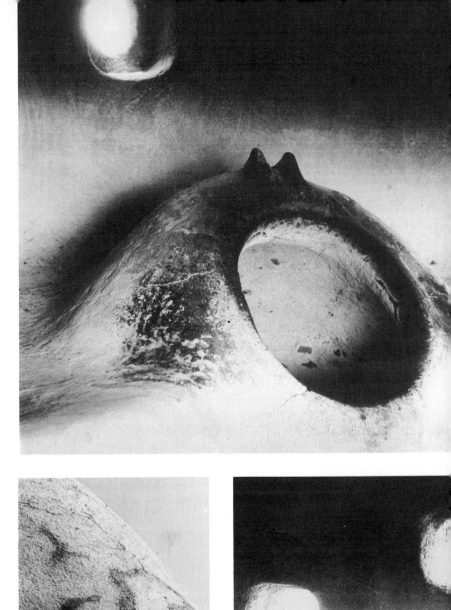

The beehive villages are a world almost exclusively composed of molded, whitewashed mud. Mud walls sweep into seats and fireplaces, storage bins and shelves. The bread oven shown in the photograph on the right rises like a bubble of dough from a waist-high hearth. The fire is kindled underneath, and the flat disks of dough are pressed to the upper inside walls of the oven where they stick until they are baked. From this oven come the familiar round envelopes of Syrian bread.

The builders who work with this soft material use it to its full measure and seem to have gained an awareness of its sculptural qualities. The results transcend the confines of the workman fulfilling his task; drainspouts (bottom left) and windows (bottom right) are needed, but the way in which they are handled seems to indicate that the builders took pleasure in their craft.

44

Though the structures sometimes appear capricious, they are
the solemn resolution of the natural inclination of the material
following the lines of stress, the
needs of the family to surround
and shelter a space and the individuality of the builders' hands.
There is little that is arbitrary in
indigenous people's lives; decisions are dictated by tradition,
the pressures of environment and
the economy of materials. Pictured here is a row of animal
stables; the small dome is a
chicken coop.

If you slice a dome in half through the center and examine the resulting cross section, you will see it is an arch. A dome, in other words, is an infinite number of arches rotated around a center point. Both forms allow materials that are relatively small in size, like stones, or of soft composition, like mud, to span horizontal distances: the dome and the arch are held together by the force of gravity pushing or compressing the material tight on itself, as a stone wall is stacked up against gravity. A third rounded roof-form is the barrel-vault structure, which again is a series of arches, this time stacked side by side. Vault-roofed structures are found all over the world, usually in areas where there is a lack of wood or other stiffening material. The vaults in these photographs were built by semi-nomadic Bedouins, who settle in an area for a few years and then move on. They are used for both grain storage and housing. Whole vault-roofed villages can be found in many parts of the Tunisian and Libyan deserts; they are built mostly of loose rubble plastered together with clay.

There is a lyrical decree in the art of building, an understanding, waiting to be used by builders with a sensitive hand. It may be an approach to the environment or just a light touch, or the way the parts combine.

Indigneous architects of northwestern Bulgaria have found this approach by necessity and are not aware of the subtleties of their work. They build with loose pieces of stone, fitting each on another in a tapering wall, which is thick at the bottom, thin at the top. They bevel and butt, lap and clasp; for every angle and curve in every stone, there is a matching one in another. The stones are held in place only by the friction of their own weight. Resting on top of the walls is the roof structure of small unsquared tree trunks. There are notches in the walls for the roof beams, which are held there by the angle of the roof and the weight from above.

On top of the rafters are light reeds which are used for insulation and to form a base and cushion for the tiles that hold them down. The roof tiles curve and interlock on each other. The whole building is fitted, notched and locked in place without mortar, nails or fastenings of any kind. It is the perfect example of assembly by form, not device. It is also an architectural sandwich, with compressive, heavy stone on the bottom, light, tensional wood and reeds in the center and heavy tile on top. The stone and tile are hard shells to face the weather; the soft underside of reeds and wood is protected to keep the structure solid.

In the east of Europe, from the Black Sea to the Carpathians, from the Danube to the Dniester, the landscape is covered with a thick tapestry of spruce, fir and some hardwood. Wood is almost the only building material. A woodworker in one of these forests finds a pleasing selection before him. He can control and regulate strength, weight and structure by his selection. He may choose as many as five different kinds of wood for one chair: black walnut for the rockers, because the nap on the surface will help to keep them from sliding on the floor; powerful ash for the legs; hard deep-grained cherry for the seat, strong, flexible hickory for the back and arms; and dowels and wedges of tough birch to peg the chair together. Indigenous builders have found that a tree is not uniform throughout. The gnarled bird's-eye grain of the wood at the crotch, where the tree has built up strength to support a long limb, is very hard; the outside of the trunk is soft and pliable; the inside heart wood, the column that the tree rests on, is strong; the sap-rich roots, durable and resistant to rot, are flexible like vines.

From year to year the tree varies as the climate changes and the forest conditions alter. Some years the growth is thick and generous and soft to the chisel. Difficult years add thin rings of hard dense wood.

From the first cutting to the final stages of decay, wood retains its appeal. New sapwood is heavy and moist, the grain fat with resin. As the sap dries, the wood becomes light and workable and warm to the touch. This is the time when woodworkers take great pleasure in working their material. As time passes, worked wood becomes dark and oily and smooth from years of handling, while wood left to the weather turns craggy. When wood is old, it develops the intricate and complicated patterns of decay. Wood is perhaps the material closest to man's own temperament—infinite in its variety, vital and filled with imperfections. Each species, each tree, each limb, each trunk is an individual and should be so treated. The woodworker adjusts his pace to the individual, at times asserting his strength, at times following the needs of his materials.

The rural people of Eastern Europe maintain a very stable society. Property remains in a family through many generations; a house is built to last for hundreds of years. Because of this long-term thinking, few houses are built, but each becomes a landmark.

Walls are made of heavy timbers, notched and mortised together. These massive beams move on their joints with changes of weather, expanding when it is humid and contracting when it turns dry. To permit this movement, the joints are pinned with wood instead of being fastened with iron. This allows the timbers to change without being torn apart. The wooden pin is forced into the hole and keyed with a wedge to hold it in place.

Builders know that wood will warp, sag and lose some of its rigidity over a long period of time, so they build to accommodate these changes. A fifty-foot oak beam will settle eight inches in a hundred and fifty years, so they set an angular stanchion to take hold when the time comes.

When oak is exposed to weather for years, the grain is etched by erosion of the softer wood. See photograph at left. Raw, unsurfaced wood lasts an amazingly long time, in many conditions longer than painted wood. Paint in poor condition may hold water under its surface, and the water rots the wood. Removing the paint can do more damage than the weather's action. If raw wood is in contact with humans or animals, the natural animal oils seep in to fill the spaces left by the dried resin.

The photograph on the top right shows the seat of a three-legged milking stool, hurriedly and crudely made a long time ago. But though it has been battered by use and worms have fed on it, time has only enhanced its beauty. Years of handling, oil from hands and drops of raw milk have polished and penetrated the grain to give it a velvet surface over the dents and bruises. In the center is a hand hole; the three circles are the ends of the legs. The stool could easily have been replaced with a new one, but the owner, a farmer in the Austrian Alps, placed a high value on it.

These handmade bucksaws (above) in an Austrian sawmill use the inherent spring of the wood to hold the saw band tight, like an archer's bow.

A satisfying balance is achieved when an undesirable element can be transformed into something useful. When Eastern European farmers clear a derelict field of new growth, the just-cut saplings are trimmed and bent into a woven fence. These fences keep both small and large animals in or out and also use very few nails, which are often hard to obtain.

In parts of Bulgaria and Rumania, branch ends woven together become fencing for corn cribs, chicken coops and yards. Twigs are used as light rakes. Woodsmen, farmers and house builders consider the whole tree as a product to be used—not just the tree trunk for lumber.

In the powerful wooden houses of Eastern Europe, huge timbers are piled on one another in majestic decay. Inside, generations of the family come and go.

Over the house is a great deep roof. Under its ponderous ribs, near the long slanted peak, on winter nights the children are sent up three stories between joists and rafters to the loft, to sleep in warm protection. Through the years the family and the house become one: each reflects the other's personality. The family maintains and patronizes the old house, and the house protects and solidifies the family.

~

The small northern Rumanian towns are dominated by huge brooding roofs, lined up on winding dirt roads. Even the fences and gates are roofed. Within, there is a great feeling of protection; the roof is heavy, the walls thick. The houses, in ancient tradition, are closed to the outside. All the life and activity are on the inside.

As the smooth rounded form spoke for the vernacular of clay in the Syrian beehive villages, sharp corners, steep angles and flat walls speak for the vernacular of a wooden village. The wood expresses itself just as clearly as the clay did. This village is in the Maremures section of Rumania on the Russian border.

Beneath the great roofs, the houses are often small. There are few windows and doors, and the roof overhangs. As a result, the inside is dark, but it remains warm during the long cold winters. Even today most rural Rumanian villages are without electricity; they use homemade candles and kerosene for light, the central fireplace for heat.

The walls of the houses are made of squared solid timbers, six to eight inches thick and some considerably more. Windows are difficult to install and glass is expensive, so they are small and few. Parchment windows can still be found in some old houses. All the intersections are held together with notching and pinned with wooden nails. On the outside of the window is a wooden grating.

Barns, as in the photograph above, are built of the same kind of timbering as the houses, but the corners cross and protrude. The proportion and design of these buildings are magnificent. Though the roofs are massive, especially over the barns, they never appear too large for the structure beneath. Sometimes the roofs extend almost to the ground to cover a storage shed or coop.

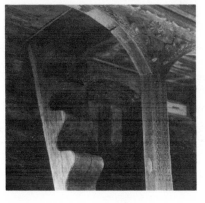

Small wooden churches reveal the Rumanian builders' art at its finest. The eaves of this little church barely allow a tall man to pass beneath them, but from their lowest point they mount to a ridge five times the height of the walls, and over that the church tower climbs more than twice that height again as a conical roof itself. Builders in Eastern Europe have taken advantage of the flexible quality of shingle construction. They bend the courses around the roof and over windows and vent openings, as seen in the photograph at the right. Each of these shingles is separately carved and fitted.

The roof overhang of the church can be seen extending past the supporting posts in the photograph (top right) looking under the front eaves. To the left of the entranceway is the ladder to the steeple, carved from a single piece of wood.

When the indigenous builder is faced with the problem of joining large structural members in the simplest, most effective way, the results are usually visually satisfying. One way of approaching the problem of structuring a post and lintel is seen in the photograph on the left. The weight of the horizontal beam rests squarely on the vertical post. These two join each other at right angles, and in a square center section the post projects into the lintel to spread the weight and create a section for fastening, which is accomplished with the wooden nails. The two angle braces hold the two beams at right angles and keep the building from racking and twisting. They also absorb the vertical weight. They thrust their weight down and hold the center beam like a pair of hands cupped and squeezed together. In terms of form and stress, the four members become as nearly a whole as possible.

The photograph on the right shows the intersection of the corner of a house with a banister rail and a gate. The rail joins and becomes part of the house; it is very clear how it is supported. The hinges of the gate are tapered into

a point that is driven into a hole in the house; the other end is rounded to fit the two pivoting points of the gate. The slats of the gate project through the end posts and are pinned to hold them in place. There can be no question of how it stays together or what its function is.

Craftsmen often split the wooden pin that holds two structural members together and then drive a wedge into it to keep the pin in place. This is necessary with smaller items that are flexed and moved about, like a plow or a chair. The wedge, the pin and the beam come to fit one another in a comfortable way; where one comes out, the other goes in, and years only make the union more solid. This is because the three

are compatible; one surface of wood adapts to another surface of wood. Metal nails or screws often loosen in their holes as the wood slowly wears away from the slight movement of the harder material. Wood and metal are not compatible.

The photograph below shows the intersection of two members of a wooden plow. The circle is the end of a pin that holds them together. The end of a wedge can be seen cutting across the pin to expand it in its hole and so hold it in place.

Many different forms of fitting have been used to hold one piece of wood to the next. Usually it is done with some sort of notching at the ends where they join, and usually indigenous craftsmen are able to accomplish this without a fastening device such as nails or screws.

Some structures in the low country of Banat near the Danube River, like this small house (shown at right) are made of hardwood planks more than three feet wide and five inches thick. This timbering is so massive that fewer than twelve planks can build a whole house, except for its shingle roof.

When houses are built in rural Rumania, it is a slow and careful task of love. The trees are selected, cut by hand, hauled in by animals and squared into beams with an adz. In the northeastern part of the country the beams are about six inches square. Each timber is laid in place, measured, removed, chiseled, replaced, measured again and chiseled until the fit is right, then the process is repeated with the next timber. The house is fitted and notched together as a boat might be assembled. Each family is its own builder. The photographs on the left and above show a father and son-in-law working on the young man's house.

This was not their trade, but they were expected by their society to be able to build a house. Also shown are some of the tools used in house building.

In some parts of Eastern Europe the solid timber construction is covered with a stucco mixture. Sapling branches split in half are tacked over the timbered wall, and plaster is worked into the inside, with clay or stucco outside. The result is almost twelve inches of solid insulation.

This house had just been completed by an old man and his wife to replace a house that had been in the family for six generations and had fallen to ruin. The new house, here shown awaiting its outside covering, was built, interestingly enough, only a few feet from where the old one stood.

Beneath the bark of a tree enormous forces are set in action. In dynamic equilibrium, gravity is trying to pull the tree down, and it is resisting. When great limbs reach for distances in horizontal positions, then send twig and leaf further than wind, gravity and rain can play upon, strong powers are multiplied by distance. Small sections of wood can support these tons because of the structure of wood. When the stress is at both ends of the grain, the wood will resist; however, one ax blow aligned with the grain can split the limb in half.

Organic builders must understand the forces at work in a natural building material and the way it resists or yields. Within the limbs of a tree can be found the four major forces that control structure and strength: torsion (twist), shear (a breaking action), tension (a pulling force) and compression (a pushing force).

When wood is used to a point near the limits of its strength, it can no longer be considered just a volume of material to be worked. The grain of a tree and the way it

is placed to resist gravity are very relevant to the woodworker. When a woodworker cuts wood to build a boat, events that molded the growth of the tree determine the cut of the wood and the structure of the boat. Every piece of wood, every form, every intersection must be justified. In the prow strength must be built up without too much weight. The grain must lie at a right angle to the force to resist shear or splitting. Every intersection is a different angle: curve compounds curve. Stresses are balanced as planking is bent and fastened on one side and then the other. Space must be left for expansion, then contraction; the wood needs room to move, it must be given space to work, but the water must be kept out. The boat must be a moving working whole, more than the sum of its parts.

As the standing tree uses its grain to deflect the crush of its own weight, boat builders use the wood to structure their boats. The grain of the boards, cut from tall, straight trees, follows the outside planking from stem to stern. Inside the hull, the forces press in from the sides;

ribs throw their weight laterally. Here the grain must be curved to follow the bend of the boat's rib. This means that the ribs must be cut from a trunk and limb that twisted and bent as they grew, so that they will now hold firm in these obtuse positions.

Trees for the boatyard are selected with intense care as to their shape. Limbs are cut with the crotch left intact; these will form the deep angle of the prow when ripsawed into planks. The grain follows the U's and V's of the hull. Each section of trunk is carefully considered for placement in the framework of the boat.

~

The strange-looking planks in the photograph above were in a boatyard in Turkey, waiting to be cut into braces to clench the keel of a small fishing boat. The ordinary approach to this form of structure is to assemble three separate pieces to form a V-angle. They are glued and/or fastened with metal in the form of screws and nails to hold them together. This kind of joining not only takes longer to perform but is far less strong than the single unit formed by the tree.

In an industrial society, forests are planted and regulated to grow trees with long straight trunks. The machine cannot use twisted wood and bent limbs, for they do not conform to the geometry of duplication and standardization. The organic builder, on the other hand, not only bends to the requirements of the tree but finds advantage in so doing; it would be difficult for his own workmanship to equal its shape and durability.

The limbs of this ancient chestnut tree have grown enormous as they lean to position themselves to get the maximum light. Gravity pulls on the upper limb—which throws the lower portion of the limb into torsion or twist.

As gravity pulls down, the wood is being both torn apart and pushed together. When the limbs yield to gravity, they break off near their base, where the most force is exerted. If the break is examined, the wood will be found to be torn apart at the top of the break and compressed at the bottom.

Shipbuilders in Egypt and other countries with limited means follow the natural form of the tree in their ships. When the wood comes into the yard, it has already been worked by the tree. The builders continue the process by cutting and reassembling the bends to form the boat. Before a tree is selected, the size and curves the finished boat will have are considered, for from the curving branches and bent trunk will come the curving ribs and keel. The tree and the boat must be worked together.

The cut sections are then smoothed of loose bark and trimmed of burls. This is all done with a hand adz.

With his apprentice at his side and his simple tools, the lofter begins his work. Learning to loft the cuts for each plank, which determines the eventual shape of the boat, takes years of experience. Patiently the lofter judges each log—its grain, thickness, width, knots, bend and where it will fit into the form of the boat. He marks his decisions with chalk lines. From the center comes a long rib for midship; the sides of the log yield shorter braces and stanchions.

A straight mark is laid across the back of the bent log with a chalk line, held tight and snapped to transfer the chalk to the wood. This process is repeated until one side of the log is divided by several lines, the space between each representing the thickness of a plank. Cut lines are made across the ends. A plumb bob is used to keep the alignment true from one surface to the next. A sight is taken on the first surface and transferred to the next one, and finally the lines are continued on the underside.

The marked log is hoisted up on movable braces (photograph at top left) for the work of the two-man ripsaw (photographs at right). Each cut of the ripsaw is a slow and exacting labor; continual attention must be paid to the course of the saw, for it can easily wander from the line, which must be on both sides of the log for both men on the saw to follow. The braces are replaced and readjusted for each cut.

Often, sections of limbs are left intact and the cut is made lengthwise through the crotch of the tree. This provides V-shaped planks to be fitted to the keel section of the boat. The men in the photograph on the left had just finished their cut and were preparing for the next. The photographs above and on the right show the cut and stacked lumber awaiting further sectioning and placement in the boat. The thinner planks will be used for smaller sections and those parts of the boat that require less strength; the thick ribs will be down deep in the hull.

The photograph at left shows the completed boat structure without its covering of planks. At completion, the boat is a tree again, taken apart and reassembled in a different way. There are the same curves and bends in a different juxtaposition.

The photograph above shows planks in a boatyard in the Calabria section of Italy waiting to be cut into ribs in another boatyard, with the shape of the ribs already penciled on their surfaces. This is better than heating, steaming and bending wood into this shape, for these planks have grown cell by cell into this form and are prepared to take stress from either the inside or the outside of the curve. Wood bent into a form under pressure and steam remains bent to various degrees when cooled, but the cells of the bend are stretched unnaturally on the outside and compressed on the inside, causing the wood to lose strength, especially in tight bends.

When an indigenous society's environment dictates an economy, the builders within that society must respond, whereas in the economic freedom of an industrial society, the needs of one environment are filled by another regardless of the distance. Yet it appears that if the material goods of a society are abundant and cheap, their value to their owners disintegrates and waste ensues. The indigenous societies of the world gear their lives to a small assortment of deeply loved goods, gently made, carefully used and lovingly repaired.

Energy and effort are also commodities to be conserved. The indigenous builder has limited power to accomplish his ends and wastes as little effort as possible. Therefore, he respects those things he can find that require a minimum of his effort to alter.

~

One of the intangible benefits of indigenous technology has to do with the relationships it creates between the human being and the natural world — the spiritual satisfaction of finding an ally among alien organisms, and the separate small surprises of working with something individual. It is also gratifying to pull ends together in a cycle, to find logic in chaos and room for one's own personality in one's work. What is the necessary shape of an object to seat a human body? A successful answer need not be a familiar form. A broken tree that has survived by contorting itself to accommodate its circumstances can become, to the organic imagination, a chair (see photograph on the next page).

~

The simple expedient of angling several forked limbs into the corner of a building to hold the beams in place is the manifestation of a complex set of tacit rules and values which are completely understood by indigenous societies. The rules are born of logic and involve conservation of energy and material — an organic technology.

The crotch between a tree trunk and its branch has long been known as a far more satisfactory intersection than any that man can

make, for the wood at that point is a whole, not two separate pieces joined by mechanical means. Shown on the right is the corner of a reindeer shelter built by the Lapps in the northern part of Finland.

Until recently, Bulgarian farmers handmade almost all their tools, including their earth-working implements. One method was to cut trees and split them with the branches still on, then lash them together to form a harrow. Compared to a steel-disk harrow, this implement is short-lived and clumsy. But it is precisely its limitations that make it worthy of investigation. The design is stretched to the limit of its abilities, and so must be carefully considered by its maker and user. Given the limitations of the farmer, his soil and his available material, the harrow is a jewel of sophistication.

The forked branch is probably the most universal form among the indigenous societies of the world. If a region has any branched trees or shrubs, the men of that region will find uses for this form. These forked packsaddles were photographed in Algeria.

Cut as if to form a thumb, the sickle handle in the photograph at right was made from the small trunk of a hardwood sapling. The branch stub was left intact to form a brace and to prevent the worker's hand from riding up on the blade. Even the smallest branches can be useful in delicate tools, as, for example, the batter mixer in the photograph below, cut from the top of a small spruce tree; the five prongs formed one of the last whorls of branches on the tree.

Some shipwrights search the forest for a tree that has a branch of the right angle and thickness adjacent to a trunk of adequate length and breadth. From these they fashion a timber that will become the gunnel from midship to bow of a hardwood fishing boat. It will be bent into place, with a hollow smoothed out behind the cut branch. The oarlock (shown above) in a Norwegian fishing boat is an integral part of the boat—a supremely simple and extremely strong design.

The large as well as the small are used. A farmer in Rumania closed his fence with this counterbalanced gate made from a huge tree section. One of the double trunk sections was cut to position the tree on a pivot. The bole forms the counterbalance.

Haying requires a tool to lift large clusters of dry hay into the wagon, a tool that will slip under and go through the mound, that will be light enough to use for a day's work but strong enough to hold a heavy bundle. It must pierce the hay easily and allow it to slip off just as easily. Again, the branched fork of a tree provides suitable material and shape. On occasion the tree also provides an extra hook between the two prongs to keep the hay from falling backwards. The natural hay fork (on facing page) is still in use in rural Bulgaria.

A hayer with strong arms can lift more hay than a two-pronged fork can hold.

Hardwood trees rarely grow branches that fork in more than two divisions. Therefore, farmers and woodworkers designed their own fork, which echoes the original tree form and has a fourth prong on top to keep the hay from sliding back when it is lifted.

The blacksmith introduced iron into the farmer's life. Iron becomes plastic under the white heat of the forge fire and can be bent into many forms. In a long evolutionary series, wood gradually gave way to iron in tools. The first farmer-made implements were wooden; as iron became more available, the edges that were subject to harsh wear were covered with straps of iron, which later were replaced by solid iron. The process continued until only the handle, the part of the tool that came in direct contact with the worker, was wood. If the hand is considered the first part of the tool, the material moves through a transition from the softness of flesh to the harder structure of wood to the strength of iron.

A farmer in the east of Finland will search through a spruce and birch forest for building material. In the limbs of a stout birch he will find his scythe handle. The gentle bend of rise and fall will be cut from the branch and taken home. The bark will be stripped, and the unwanted material planed away. Then, carefully, with his knife, he will work into the curve, altering it, shaping it, flattening it to embrace his shoulder and movement as the other end works the grass. Each scythe he makes is an improvement over the last, and it follows his change in action as he grows older. As he carves, he swings the new and old scythes to feel where the placement of the handle's curve and hold should be. The handgrip is placed, the length is altered, the cutting blade and angle are determined.

One of the most exacting and important tools a farmer can make is his plow. It must be designed to fit the particular region in which it is used. Often the land condition will vary from one slope to the next, or even within a farmer's own acreage. The farmers on the southern shores of the Mediterranean, for example, work with a light, permeable, sandy soil. Their concern is to conserve moisture by freeing the surface from the ground underneath and allowing it to pow-

der, which reduces evaporation. This method is called dry farming. The photograph on the left shows an unlikely draft team pulling a farmer-made plow on the western slopes of the Atlas Mountains in Morocco, where the soil is also light and sandy.

The plow in the photograph on the right was built to be used in the light sandy soils of Syria. The pointed end, the plowshare, is designed to contact the ground at the exact angle for the existing soil of the region. If the angle is too flat, the plow will ride ineffectually on top of the surface or cut too shallow a furrow; if too deep an angle is used, the plow will dig in and will exhaust the draft animals. The plow is made of wood, held together with iron straps and wooden pegs. The hardest wood available is used for the share. The draft beam, or section running up from the share, is made of a curved limb, as is the next long section, the draft pole. One of the great problems with a wooden plowshare is its short life. Many methods of extending it have been tried, including embedding pebbles in the share and stock.

angled pieces on each side, called ground-wrests, open the furrow by throwing the soil wide on each side (see bottom photograph). By tilting the plow to one side, the farmer can turn over the fresh sod as it is moved to the side.

This particular plow was made of lemonwood, which has a thick oil that allows it to slide more easily through the soil.

To the north of the Mediterranean, where the clay-bearing soils are very cohesive, plows are totally different in form, no longer spears or arrows but wedges. The plow pictured on the right was used until recently in Rumania. In the arid south, water must be conserved; in the north, the concern is adequate drainage and aeration of the soil. The furrows are wider and the ridges higher. The plow cannot be angled steeply into the soil, for the pull would be too great. The ridges of soil are not loose strips piled up next to the furrow, as they are in the Mediterranean regions, but tightly packed clumps, and the long furrow develops a continuous ribbon of turned-over sod. The wedges that followed the plowshare are called moldboards.

As soils change, so do plow forms. The plow in the top photograph was designed to be used in the soils of Sicily, which are heavier than the soils of Syria. It is an arrow rather than a spear. The two

The moldboard plow reached a high point of perfection in Early American plows like the one shown in these photographs, which dates from 1700. On this plow the share is a separate piece of heavy iron plate; the moldboard is wood faced with iron straps. The moldboard plow performs three separate actions: the share cuts into the soil and lifts the sod up above ground; the sod is passed to the moldboard, which moves it to the side, where it is turned over by the curved upper corner of the moldboard and laid on the ground next to the furrow. The plow is light and strong because it is a three-dimensional truss with angle-bracing to follow the stress. The iron pin connecting the draft beam with the share is in tension and holds the front of the structure together. Note the farmer's use of a leather washer as a buffer between the iron pin and the wood on the top of the plow.

Animal-drawn plows varied from country to country and even from one farm to the next. They had to be coordinated with the type of soil, the moisture content, the kind of planting, the draft animals used and the farmer's own movement. When animals were replaced by mechanical power, plows were simplified, the differences between them became less, and the variations between plows used in some areas disappeared completely. With mechanical power, it was no longer necessary to design each plow individually. General plows were designed to work almost any kind of soil. The farmer was freed from the constraints of his environment. He no longer had to understand his plow in relation to his soil composition.

113

In indigenous societies, the artisan manufactures those things that the people themselves don't make. Shown are wooden grain shovels made by an Egyptian woodworker. The shovels are made from a simple piece of wood. The indigenous builder is not averse to decoration, but the decoration is usually an integral part of the design, not an addition to it. The fluting on the back end of the shovel and the ridges are both decorative and functional. The fluting enables the artisan to get two shovels from one piece of wood. By alternating the position of the shovels, as shown in the photograph above, the end of the handle of one fits into the fluting of the other. The ridges strengthen the shovel.

The milking stool accommodates both the human form and the shape of the wood. The curves of the tree fit the complex curves of the human body. The stool is comfortable and sturdy; the three legs splay out for stability and are firm on uneven ground. The beauty lies in its simplicity—design stripped for expediency.

Pictured on the left is a combination brace to hold and a roller to move a small Italian fishing boat. Simple as it is, every curve and angle is a response to its use. The top and bottom are fanned out to spread the weight against the sand and boat bottom; one end is angled to accommodate the corresponding angle of the boat; the center is curved in to cradle the keel and keep it from sliding off when the boat is moving up or down the beach.

Furniture built by indigenous people is usually heavy and massive, but it is also scarce and so must be moved from room to room to fill several jobs. The table pictured on the left was made by a Scandinavian farmer, can be disassembled into six parts, including the two wedges that hold the whole assembly together.

~

Some forms respond primarily to the material from which they are made and only secondarily to their usage. Small flexible tree trunks and limbs are used by the barrelmaker as hoops. Green hardwood sections are split in half, then stretched tightly about the barrel and turned under. As the saplings dry, they contract and grow stiff, and the barrel staves are compressed together. By contrast, iron hoops cut into the wood when it expands and loosen when it contracts. A wooden barrel is a classic of functional form, built so that its pieces become more compressed as they expand with moisture, and therefore leak less. Its construction is similar to that of a boat turned

inside out. The internal ribs on a boat function like the outside hoops on a barrel; the water is on the inside of a barrel and the outside of a boat. The important exception is that the barrel is free to follow its own form, controlled only by the shape of the wood; the boat is also exposed to the rigid discipline of a form moving through water.

Implements that take their shape directly from their working contact with the substance they are made to deal with often express their purpose in simple and beautiful terms. In the strong hands of a blacksmith, iron is given a vitality that often transcends its use. Hay forks are light, long and yellow like the hay they handle; sawdust shovels are big and round like the billowing mounds of sawdust they move (see photograph above). Auger bits resemble the soft twist of wood they cut; tools that chip and chisel wood are more angular and direct; farm tools that fight a rocky environment are jagged and sharp.

Tools that move through water or churn a liquid become soft and round. The shape of the oars used in a Norwegian fishing boat (see photograph on facing page) expresses the flow of water passing by, firm yet yielding. The builder knew the pull of a working oarsman. He designed the oars so that they move freely through the water while the wide sections provide the leverage with less speed and a heavy push.

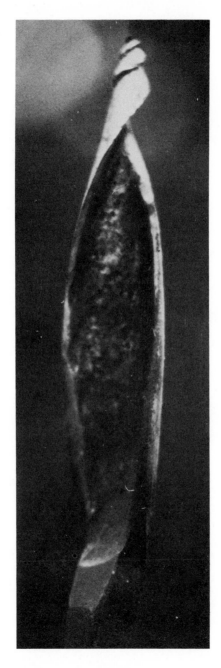

The heavy-forged, bayonet-pointed implements pictured on the facing page were made by a blacksmith on the black and red lava fields of Mount Etna in Sicily. In this hostile environment, the shovel has become a weapon of war to fight the rock and crust of the lethal mountain.

This carpenter's auger from Austria (see photograph at left) and locking-hook (above) to hold a well bucket from Bulgaria are small simple tools.

When four fingers of a hand curve to one side and the thumb lies opposite, the palm is a gentle valley, a soft pocket to receive the surface of a tool and propel it on its course. Fingers and thumb close to guide it and hold it in place. The palm plane in the photograph on the left was designed by Syrian handle-makers to smooth shovel handles. Its small blade cuts long wafers of wood as it travels the length of the shaft. The other end of the blade rests between hand and thumb, which help direct the tool. The rounded triangle of the plane was designed to be an intermediary between the wood to be smoothed and the hand.

~

From the moment the builder puts away his hammer, the destructive forces of nature start decomposing his just completed structure—the wind erodes, the sun dries, the rains leach and wash; insects, animals, organisms of decay eat at it; and time itself pulls it apart. The fragile things he builds must be sheltered, continually maintained or left to lead

their short lives. The structures that endure contain material that resists these forces, or so much bulk that a great quantity of natural forces is required to consume them.

The roof is usually the first part of the house to wear out; it receives the brunt of the weather. The indigenous builder has tried in many ways to build durable, insulating, weatherproof, maintenance-free roofs. The thatched

straw roof, like the one in the photograph above, which was taken in Bulgaria, is used in many parts of the world. It takes skill and patience to put together the bundled and clipped thatch, but it will last upwards of eighty years if the climate is not too moist. The small forms below in the photograph, which reflect the roof's shape, are beehives made from branches split open and woven.

This unkempt shaggy mound is another kind of straw roof, which used to be popular in rural Rumania. The house is built half underground and half into its haystack roof. This kind of roof is far easier to build than the thatched roof but provides less insulation because the thickness is not controllable.

Stone is another indigenous roofing material. The photograph above shows a random-pattern stone roof in Norway. There are various methods for holding the stones in place, but the most common is to drill and wire them to the wood beneath. This means that it is very difficult to find the origin of a leak because of the many little seams and intersections. But if leaks can be prevented, the roof will stay intact until the timbers below it rot.

Another Scandinavian method of roofing is the trough and peak found in Finland (see photograph at left). Hollowed logs are turned one on another to form a unit that is not only weathertight but extremely enduring.

One of the most pleasing accomplishments of the indigenous builder is self-regulating systems. These require less work to build and maintain, and provide the satisfaction of creating a living system. The builder treats his house as an organism that he brings to life, which then, by its own mechanisms, sustains its functions by renewing itself.

For years the Norwegians have used organic roofs. These begin with a layer of birch bark as a sealer. Overlapping birch bark is weathertight and almost completely rot-resistant. Over the bark, squares of pasture sod about five inches thick, with roots and grass, are laid. As the seasons pass, the sod perpetuates itself; root intertwines root, and the roof becomes a solid whole which rain and weather only strengthen. In the winter the dead stalks of grass hold the snow for effective insulation. The spring rains beat the grasses down, so that they shed the excess water, then bring the roof to life again. In summer grasses grow long and effectively reflect the sun's heat.

As might be guessed, the sod roof is very heavy, especially when saturated with water, so the structure beneath must have a carrying capacity several times that required by a wood-shingle roof.

Three boathouses on a Norwegian fjord. The local people were unable to say just how old they were but said they were definitely over a hundred years and possibly two to three hundred years. To the best of their knowledge, nothing had been done to the roofs in at least a lifetime, maybe two. The roofs are still tight to the weather. Visually, the houses blend into the forests and pasture; they actually become part of their environment, although when they were built, this was secondary to the expedient of creating a solid practical structure that would last.

Sometimes masons take the same approach. They lay stone walls with mud or clay instead of inorganic mortar. This provides a footing for vines, which, when they reach maturity, form a solid structure that if coaxed to grow in the right direction, holds the stone in place and is tight enough to resist the weather.

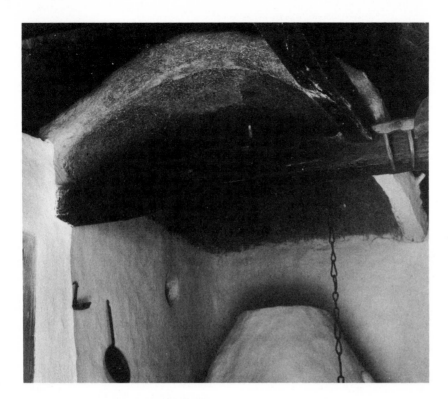

Houses with deep, pointed roofs of straw also act as organic units in the forests of Rumania (see photograph at right). Although they do not grow, they maintain themselves. They are built without chimneys to avoid letting out and so wasting the smoke. The stoves and fireplaces empty the smoke into the underside of the roof. The rafters fill with smoke down to the ceiling of the living room, and as it slowly filters out through the thatch it provides central heating, cures the meats that are hung in the peak, and at the same time pre-serves the straw by checking rot and decay. The act of living in the house preserves it.

Inside the chimneyless thatched house, the stove, the top of which is seen in the lower part of the photograph above, empties its smoke into the clay shield, which allows it to spill upward to fill the roof. Because of the porous nature of the thatch, the smoke sifts out before it drifts down to bother the in-habitants. This kitchen had not been used since it had been re-built and painted; normally the shield is smoke-blackened.

If a thatched roof is left unattended for a generation or two, the straw begins to rot. But if the climate and prevailing conditions are favorable, it will begin to come to life again and reestablish itself.

The moist, decayed straw becomes an organic growing base that is very suitable for mosses. As the straw gives way, the mosses spread and soon cover the entire roof. With a roof that pitches steeply, the moisture is held out and the roof is virtually self-perpetuating.

The darker areas on the roof of the Bulgarian barn in the photograph on the facing page are patches of moss; the structure to the left is a woven corncrib.

~

Self-regulation in buildings can be achieved in many different ways. The rugged wooden houses of northwest Rumania in the photographs on the right and on the following pages are examples of materials providing the regulatory device. These houses are built on the rock ledges that jut through the forest in the Transylvanian mountains. In the winter the north wind cascades down from the plains of Russia; in the summer there are the hot winds off the Yugoslavian plateau. The winter winds are heavy with moisture, the summer winds are very dry.

The people cover their roofs, and sometimes their whole houses, with shingles cut from a type of fir that grows in the area. This fir is extremely sensitive to humidity changes; it expands rapidly in moisture and contracts in arid air.

The shingles are notched and grooved to fit into one another (see photograph below). In the winter, when the moist winds press down on them, the shingles turn dark and thicken with water. Row on row, they grow fast together; the more the wind torments them, the more solid they become. The house shuts itself in for the winter, secure against the cold. As the weather moderates, the drying wind from the south begins. By early summer the shingles have turned light gray and relaxed their shoulder-to-shoulder press. August opens up the house. The dry shingles move aside, allowing the air to run through the house and cool it.

Another kind of self-regulation comes from the form used and the limits of size. Most of rural Egypt is spotted with large pigeon cotes. Pigeons provide both fertilizer and food. Each of the hundred-odd holes in the tower, which extend through the structure, provides an ingenious brooding nest which accommodates precisely enough young to maintain the rookery. The unfortunate extras, usually the smaller ones, fall inside the tower, to the bottom. All that the owner has to do is enter the base of the tower and pick up his daily harvest of squab. Because it is self-regulating and self-maintaining, the cote tends to support a larger, healthier stock.

In yet another type of internal control, a bad situation is used to correct its own evil. In central Morocco, the Atlas Mountains taper into a broad coastal shelf above the Atlantic Ocean. Here, a satisfactory balance has evolved between building and farming. With care, the soil can be coaxed into producing a fair food crop, but it supports little vegetation substantial enough to be used for construction. Loose stone, however, lies thick and scattered through the black loam. A curse to the farmer, it must be cleared to till the soil. But the farmer and the builder are the same man, so the stone is a boon as well as a bane. Carefully clearing the white rocks from his black soil, the farmer piles them up. Later he washes the dirt from them and fits them for a house wall (see photograph at bottom right).

For as many years as the Egyptians have been drinking the water of the Nile, it has been muddy. And for as many years they have used the large vessels with rounded bottoms called amphorae (see photograph at bottom left) that are placed in stands and filled with warm brown river water. This water slowly seeps through the vat, a process that cools and filters it, and drips into another container beneath. Amphorae, made with mud from the river, are used to clean river water; the problem carries with it its own solution.

~

When Bulgarian farmers clear an area of trees to make a new field, they lessen their work by leaving some trees spotted through the new field. These are stripped of their branches, except for the lower ones and those at the very top. At haying time, these trees provide poles for the stack to build on and a platform above the damp ground and cows (see photograph at right). The trees remain healthy and growing, and as they grow, instead of gaining more height, the lower branches give a wider platform.

artisans in an indigenous society may specialize in working on one kind of material, like woodworkers or blacksmiths, or in making one product, such as a wagon, which may take more than one kind of material. Artisans know the crops, livestock, terrain, climate and people in their area and make their products respond to specific needs and conditions. Their workmanship is of high quality; the artisan works directly for the user. The farmer buys few goods, but each must be a superior product.

Although the artisan tries to make his goods as responsive to the needs of the user as possible, a certain kind of direct, spontaneous invention is eliminated when the user is no longer responsible for the production. The artisan understands his tools, his material and the product, in terms of its manufacture, more thoroughly than the home craftsman, but he responds to manufacturing needs more readily than to use. He cannot possibly foresee a necessary innovation in a tool the way the worker can who spends his days using that tool. The artisan eliminates the farmer's

autonomy. There is no longer inventive action volleyed back and forth between man, his products and his place. Change must go from the farmer through the artisan.

But even though the artisan represents the first step in removing production from the user—a process that has moved vast distances in the modern world—the artisan maintains a pristine ratio of one worker and one result.

No written plans are needed, no concern with passing information from worker to worker. The artisan's plans are in his mind; they are intricate or simple as the moment demands, change and invention are easily accommodated. The products, though preconceived, are spontaneous and varied. Each is as fresh and vital as though it were the only one ever to exist.

~

In the artisan's marketplace there is always an anticipation of the unexpected. Though the products are farm or household goods, which are sold in large

numbers, no two are alike. The
differences come from slight
errors in construction, variations
in material and, above all, because
the indigenous artisan respects
the individuality in product and
person. Pictured is an Egyptian
woodworker with his wares. His
standard manufacture is about fifty
different home and farm implements; those things that he normally doesn't manufacture he will
be happy to make. The custom-
made is common place.

146

Within the rural marketplace, there is diversity from shop to shop. Plowshares evolve to fit the soil and crops of a specific area. Each blacksmith in the area makes this type of share, but each adds his own variations.

Because indigenous craftsmen recognize the need for individuality, they often intentionally alter their product's form from one to the next. Hoe heads, as in the photograph above) are a staple, yet the blacksmith deliberately places the square holes in different positions on different hoes, bobs the corners on one and leaves them on the next. To emphasize the separateness, the small stamped patterns on the surface are changed from tool to tool. These differences indicate an awareness for the need of change. To the left can be seen a type of cultivating tool.

Individual variations can also be found in the Egyptian water jugs in the photograph on the right—small alterations in diameter, length, design and decoration. It is in the nature of production to build with static repetition, producing exact replicas piece after piece; it takes initiative, energy and effort to break out of this set.

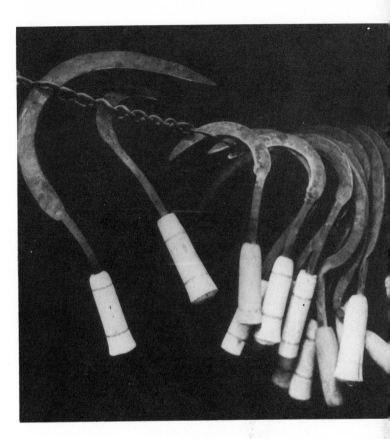

The indigenous artisan also responds to his customers' individual desires. At harvest time a farmer spends his days leveling the grain with his sickle—difficult and tiring work. He comes to know the motion well. The tool at the end of his arm, the sickle, must be shaped to his needs. Buying a new sickle is an important event. One is chosen from those already made, reheated, worked and reworked until it fits the farmer's swing perfectly.

The tools used by farmers, woodworkers, blacksmiths and most other indigenous artisans and craftsmen become very specialized. These tools are all hoes, but among them are many variations of digging, scraping, turning, separating, chopping, mulching, splitting, ditching and mounding tools. A farmer wants his tools shaped as accurately as possible for the job they have to do in the specific kind of land he tills. At the height of the artisan economy—from 1550 to 1850—

toolmakers' catalogs in England and Europe offered dozens of variations in every type of tool. One London toolmaker offered forty-five different kinds of wood-finishing chisels, each in a range of styles and sizes. This was just for a finishing chisel produced by one manufacturer. These differences sometimes may have been slight, but more important is the fact that artisans felt the need and had the involvement in their work to demand a precision in their tools.

The Syrian artisan (at left) specializes in the construction and repair of wagons and wagon wheels. But since he makes his own nuts, bolts and iron tires, he is also a blacksmith. The wagonsmith holds a high position among the artisans, for his craft is one of the most exacting.

The Syrian blacksmith above makes many kinds of iron goods at his forge, mostly small hand tools, keys and balance scales.

This young Moroccan is an apprentice blacksmith. The smith he works for is a specialist in wrought-iron fences. The smith twists the forms and cuts the frames; the apprentice assembles them. Wrought iron is an excellent material for this; it is tough, but under the fire, it will flex and can be worked into intricate shapes. Westerners have admired these fences, and in a good example of gross misunderstanding of and insensitivity to a material, have copied them in cast iron, which is hard and brittle; it shatters like pottery but is resistant to heat. Its natural form is a solid heavy mass, like a stove.

Each kind of material has its own form. Artisans come to know their materials and just which forms they assume comfortably. The photographs on these pages show a Syrian coppersmith surrounded by his work. The indigenous artisans would certainly not be able to give a description of their material with any approximation of scientific terms, and possibly might not be able to assemble the words to describe it. But it is certain that they know the materials well; they have developed an intimacy and an intuitive knowledge that goes beyond the spoken language.

for a suitable length of time and enough toughness to keep the tool from chipping or cracking if it hits a rock. Increasing one of the qualities lessens the other; the tempering balance is crucial. Tempering is achieved by reheating the finished tool to an exact amount, then quenching it in oil or water. To find the exact quench point for a proper balance, a ram's horn is held on the heated blade; if it crackles without producing smoke, it is too hot; if only small curls of smoke gather, it is too cold; if, as shown in the photograph on the left, there is an abundance of smoke, the smith rushes the blade to the quench.

In some places the blacksmith is still regarded as a somewhat fearsome person. His art is difficult to understand, his materials come from the depths of the earth, he works in a darkened recess, building and tending pits of intense heat, pulling from them white-hot irons that he works violently. But the mountain village in the Catania section of Sicily—like other rural areas of the world—has come to depend heavily on the local blacksmith.

By chance and investigation, artisans over the years develop strange little intimacies with their materials and the processes by which their products are made. In the mountains of Sicily, blacksmiths have developed a peculiar but effective method for determining the tempering heat of hatchets and axes. These tools must have a balance between enough hardness to enable the blade to fold its edge

Most blacksmiths do not work with wood; tools requiring iron working parts and wooden handles are a collaborative effort of two artisans. The customer selects a hoe, shovel or chisel from the smith and takes it to the handle-makers to have a handle fitted.

Pictured on the far left is a Moroccan handle-maker shaving a long shovel handle with a draw-knife; in the other photograph a Syrian artisan is straightening a pole for a shovel handle. Wood is scarce, so the woodworker must settle for the material available.

The rhythm of a woodworker is far removed from that of the blacksmith. The blacksmith works swiftly, with the forceful movements necessary to push the red-hot, tough iron into shape before it cools. He is urgent and aggressive. The woodworker is methodical and quiet.

Beside him lie his tools, waiting their turn: long-bladed knives to cut long thin wafers of wood, hatchets and adzes to cut chunks; curved shank gouges to cut spirals. By following the edges and flats of a tool with thumb and eye, the woodcutter can determine the character of the finished piece by his choice of tool. His tools are a vocabulary; each one possesses a subtle inflection of meaning. With his tools and the motions by which he uses them, a conversation is conducted between worker and material. The wood argues in knots and agrees in smooth grain. Patiently the Sicilian boat builder in the top photograph cuts and bevels the pieces to fit into his product. On the right is a Syrian woodworker who builds farm implements. To the left in the photograph are some of his products: plows, harrows and packsaddles.

The work of most artisans is a total involvement of mind and body. Many have found that two hands alone are not enough. They use as much of their bodies as possible. Feet may be extremely important. Shoes would be a terrible burden to the Moroccan cooper (see photograph at left). His feet hug and toes curl around the piece of wood he is working, like a highly adjustable vise.

At times, as with the Egyptian basket-maker (see photograph above), the foot and leg are used merely as a very mobile weight.

Feet and toes to the Moroccan woodturner or the Egyptian palm-frond cutter are far more than just assistants to the hands. The photographs above and on the left show the hands and feet of a man operating a hand-powered lathe. The worker's right hand swings a bow back and forth to spin the piece being cut, the right foot and left hand operate the cutting tool and the left foot braces the whole mechanism. The palm-frond cutter in the photograph on the right clenches the work between his first and second toes.

When a choice must be made between the hands and the feet, the hands usually get the skilled work and the feet the laboring jobs, as shown in this photograph of a young Moroccan sharpening a knife blade on a foot-powered grinding wheel.

The thigh is an always handy and highly adjustable worktable. Some artisans, like the Moroccan pictured on the right, use their thighs so much that they must cover their legs for protection.

Artisans also use their bodies as measuring devices. Many measuring standards come from parts of the body: the length of the first joint on the thumb, the distance from thumb to little finger on the outstretched hand, the length of the foot, the distance from armpit to fingertip, the spread of two arms. Many of the best artisans today seldom use a measurement other than the distance between two points already known to their hands and eyes. The Egyptian potter making drain tile on his wheel, as shown in these photographs, measures the length of the tile very simply by the distance that his arm can reach inside.

The exchange of goods in an indigenous marketplace is satisfying to both maker and user. Before the product reaches the hand of the user, it is something personal to both. There is a recognition that the maker is giving the user part of himself, his knowledge, his energies and a segment of his finite life, and that the product will have a place in the user's environment, his time and his hand.

In earlier times, when a farmer needed a hinged gate to close his fence from pasture to barn, he made it. He cut the standing tree and seasoned the lumber; he carved, reamed and fashioned the hinges. He installed the gate and used it until the termination of his or its life. His background had prepared him to be a generalist equipped to handle the situations that arose within his environment without help from his society. When the blacksmith moved into the community, his craft solved some of the farmer's problems more successfully in iron. Production moved the first step away from the consumer, the farmer became a little less of a generalist.

From the independent family workshops and benches behind the kitchen, from the farmer's shops in the barn, the production of goods moved to the local artisans—the wheelwrights, shipwrights, tinsmiths, coppersmiths, coopers, blacksmiths, cabinetmakers, the local market. But the community was still self-contained; it made only what it needed and used only what it made.

The cohesion of the community was broken by the use of power beyond man and his nucleus of work. Machine technology was born in wind and water mills. Its power reserves were huge, and the work it produced was limited only by the number of skilled men to keep it running.

The windmills utilized a power that has existed as long as the earth, tapped it for use and then released it to pass undiminished into the future. But wind cannot be stored, and the still days of summer stopped the windmills just as the winter ice froze the sawmills. Production couldn't wait for the climate, so the mill became the factory and the power was provided by steam—expensive and demanding, noisy and defiling, but ceaseless.

The factories led to specialized production. The factory system demanded unification of action, material, size and product; it produced identical objects by the mass.

Machine technology freed man's body, but he himself became a product of his production.

He absorbed the idiom the machine used to manufacture his comforts; he sought symmetry, uniformity, standardization.

The human body, an organism that cannot see, hear, run, climb or swim very well, is the body of a generalist, for it can do all of these things to some extent. Now it was forced into the role of a specialist. The vast complex of brain that had brought the mill and factory into being sat idly by and watched the monotonous repetition of eight highly controlled fingers and two opposing thumbs perform simplistic tasks. The abstract, individual brain came to be used for meaningful work in only a very small fraction of the human race. Most humans go through life in a mechanical society without contributing either to themselves or society. Most individual expression is relegated to the role of trivial amusement. Personal expression has become the property of the artist, the hobbyist, the collector.

The generalist in a simple society finds his satisfaction in his work; no other expression is necessary. Modern technology is concerned only with supplying the demand for goods, not with the activities that produce these goods.

The generalist was an individual because he was used to thinking for himself. He relied on his own imagination to solve his problems. Unprecedented situations, which constantly arose within his environment, demanded original thinking. The artisans were the first step away from this position. They had established ways of acting. The individual in a mass society today looks to the society for the control of his environment, and perhaps for much of his thinking. Group activity has become the path to significant action. The power of solitary rational judgment has almost disappeared.

The factor that above all others differentiates man from other species is his need to alter his environment. He is essentially alien to almost every environment. Regions are too hot or cold, too wet or too dry for him. They do not provide him with food to eat, water to drink or a place to sleep, but he cannot wait to adapt genetically, so he must plow and plant, dig and build.

Every organism changes its environment to some degree. The very fact of its existence within the environment alters it. It eats and converts foodstuffs, consumes energy, dies and leaves its body. Many species are home builders, and sometimes their homes are also their food. Trees are felled, rivers diverted, grass lands destroyed and islands created. But none alter the land as drastically as man. Most other creatures have been working their environment in the same way for countless generations, and the environment has changed over these years to accommodate their work and even to expect it. Organisms become part of their environment, and if they were eliminated, their environment would have to undergo crucial reorganization. Man has not become part of his environment, and rarely has the environment accepted him.

Most of man's drastic actions have come within the last two hundred years—a microsecond in environmental terms. It is possible that the environment will eventually come to accept us, but it will probably require many hundreds of our generations, and the environment will certainly be something quite different from what we know now. Just as possibly, man could be the germ of destruction that the environment could never assimilate, and his work would lead to the total collapse of the biosphere and, of course, man.

If Western man is to live within the biological fabric and leave it suitable to other living, he must regain a regard for conservation and economy that modern technology has taught him to disregard. It is easy to canonize indigenous societies, but to follow their ways, we would have to become more sophisticated than they, for they, like us, are only acting out of expediency. They must conserve and work within their environment to survive. Our expediencies are consumption, force and waste. If we were to follow the example of indigenous peoples, we would become the first human society to impose a purposeful self-discipline.

It is a culturally unhealthy act to look for advice to societies that chronologically should be emulating us, but it seems to be necessary to reestablish our direction. If we succeed in assimilating the ecological sophistication of indigenous societies with our knowledge and establish a modern organic technology, the result will be a happy one.

If modern man had the free thought of the generalist in an indigenous society without the latter's ponderous burden of tradition, superstition and fallacy, unprecedented results would follow. If modern man were as free to regulate his house or town as the Tunisian mountain dwellers or the Egyptian villager, the results would be far more exciting. If modern man understood the use of materials as the Syrian blacksmith understands his iron sickle, then modern materials might be used more honestly. If modern man practiced the same economy of effort and material that the Finnish farmer uses in determining his scythe handle, there might be freedom from waste. If the modern system of manufacture were reorganized to provide satisfaction for the people who produce, those people would surely begin to have a more significant life.

Yet technical manipulation alone is not the solution; it is the result, not the cause. Before modern technology becomes compassionate, modern man must.

About the Author

Christopher Williams is at present
a visiting professor of environmental design at the University of Al-
berta in Canada, on leave from the
Cleveland Institute of Art. Like his
wife, Charlotte Williams, he re-
ceived his training at the Pratt
Institute in Brooklyn, New York.
The Williamses gathered the material for this book during a fifteen-
month, 40,000-mile trip by camper
bus with their four-year-old son,
from Spain to Morocco to Egypt,
across North Africa, through Syria,
Turkey, Eastern Europe, Italy and
Austria, north to Scandinavia and
across the Arctic Circle in Finland.

In most of the modern world, machine technology has destroyed the harmonious association man once had with nature. This book focuses on those "indigenous" societies—from the Berbers of North Africa to the Lapps of northern Scandinavia—that still maintain the old organic relationship between the two. In order to survive in an often sparse environment, these people must constantly be aware of the finiteness and potentialities of nature's resources. Theirs is a closed system with a minimum of waste. This does not mean that they ignore aesthetic appeal; instead, design and function become one.

This book is both a handbook and a philosophical plea. The reader learns through text and photographs the techniques used in various indigenous communities: how to construct roofs to last for centuries, how to build ships by following the natural form of the tree, how to build a home underground using "subtractive" building and the concept of negative space. Unlike the products of modern technology, which are mass produced and allow no variation, an indigenous design is one individual's creative expression and can never be precisely duplicated. To work with and within the environment provides great spiritual satisfaction. After presenting the indigenous design techniques of many different communities, <u>Craftsmen</u> <u>of</u> <u>Necessity</u> challenges the reader to incorporate their ecological attitudes into his contemporary existence.